Material Interfaces as Energy Solutions

Topical Conference at the 2022 AIChE Annual Meeting

Phoenix, Arizona, USA
13-18 November 2022

ISBN: 978-1-7138-7914-5

Printed from e-media with permission by:

Curran Associates, Inc.
57 Morehouse Lane
Red Hook, NY 12571

Some format issues inherent in the e-media version may also appear in this print version.

AIChE
120 Wall Street, FL 23
New York, NY 10005-4020

Phone: (800) 242-4363
Fax: (203) 775-5177

www.aiche.org

Additional copies of this publication are available from:

Curran Associates, Inc.
57 Morehouse Lane
Red Hook, NY 12571 USA
Phone: 845-758-0400
Fax: 845-758-2633
Email: curran@proceedings.com
Web: www.proceedings.com

TABLE OF CONTENTS

PLASMA CATALYSIS

INTERFACIAL SYSTEMS FOR ENERGY APPLICATION: MODELING AND SIMULATION

INTERFACIAL SYSTEMS FOR ENERGY APPLICATION: EXPERIMENTAL VALIDATION

INTERFACIAL CHEMICAL CONVERSION

LOW DIMENSIONAL MATERIALS

MACHINE LEARNING IN MATERIALS DISCOVERY

220b Chemical Looping Ammonia Synthesis: Thermal, Microwave, and Plasma Approaches

Sean Brown, Brandon Robinson, Jianli Hu

Ammonia synthesis by Haber-Bosch is responsible for 1% of global energy use per year, 2.5% of global CO2 emissions per year, and for supporting approximately 4 billion additional persons' nutritional requirements [1–3]. The importance of Ammonia to the world economy is not in doubt, but ammonia may also prove to be a future energy carrier due to its high hydrogen content. The chemical looping ammonia synthesis (CLAS) reaction seeks to decouple the high energy dinitrogen cleavage reaction from the ammonia synthesis reaction. The ammonia synthesis reaction may proceed as a hydrogenation or hydrolysis reaction, depending on how the loop is constructed. Due to the passive nature of this process, nitrogen may be stored and retrieved later at the site where it will be used.

Fixed bed thermal and microwave reactors were used to evaluate the three-cycle ammonia synthesis productivity of CLAS materials; Fe and CoMo. These materials were physically mixed with the insulating microwave sorbent, SiC, in a 1:1 ratio to enhance their microwave heating capability. Microwave heated samples were found to be more productive but deactivate faster than thermally heated samples. Regeneration strategies and deactivation modes were considered and analyzed.

During thermal fixed bed experiments CoMo was found to evolve ammonia during nitridation, in the absence of hydrogen [4]. A dynamic surface, a shrinking core and counter diffusion were proposed as the source of hydroxyl ions stabilized by nitride ions, reacting at the Co_3Mo_3N surface detected with XPS [4]. The existence of an unreacted oxide core containing molybdenum bronzes, H_xMoO_3 was confirmed using Raman spectroscopy. This OH^{-1} diffuses outward as the core is stabilized by N^{-3}, resulting in the observation of gas phase H_2O and NH_3 [4].

The productivity and kinetics of common CLAS candidates was experimentally evaluated in thermal and thermal-plasma conditions. Common materials include Mn, Fe, and CoMo, as well as promoters and oxides are considered in both systems. Plasma can pre-activate the N≡N system and lower the energy required for reaction. Plasma treatment can also interact with the surface in different ways, perhaps circumventing some of the deactivation modes encountered with thermal only CLAS processes. Spent materials are characterized with SEM, BET, and XRD. Time on stream reactions of outlet gases are analyzed and OES is used to determine the species present in the plasma and evolving from the surface.

References:

(1) Erisman, J. W.; Sutton, M. A.; Galloway, J.; Kilmont, Z.; Winiwarter, W. How a Century of Ammonia Synthesis Changed the World. Nat. Geosci. 2008, 1, 636–639. https://doi.org/10.1038/ngeo325.
(2) Pfromm, P. H. Towards Sustainable Agriculture: Fossil-Free Ammonia. J Renew. Sustain. Energy 2017, 12.
(3) Appl, M. Ammonia. In Ullmann's Encyclopedia of Industrial Chemistry; Wiley-VCH Verlag GmbH andamp; Co. KGaA: Weinheim, 2006.
(4) Brown, S. W.; Jiang, C.; Wang, Q.; Caiola, A.; Hu, J. Evidence of Ammonia Synthesis by Bulk Diffusion in Cobalt Molybdenum Particles in a CLAS Process. Catalysis Communications 2022, 106438. https://doi.org/10.1016/j.catcom.2022.106438.

Haber-Bosch: 300 - 500 °C, and 20 - 100 MPa

$$3H_2 + N_2 \rightleftharpoons 2NH_3$$
$$\Delta H = -46.35\ kJ \cdot mol^{-1}$$

Chemical Looping Ammonia Synthesis

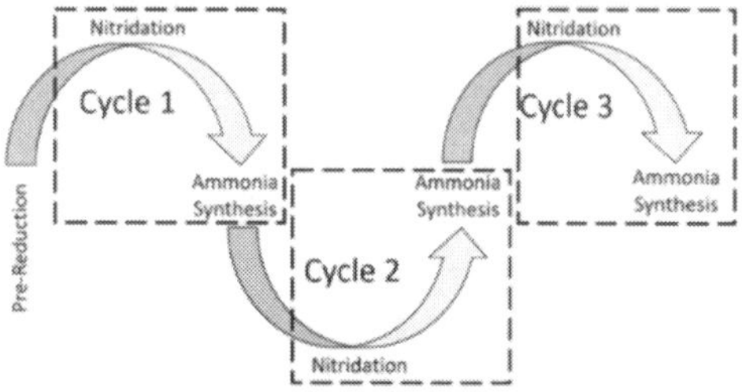

1 atmosphere - separated reaction cycle

(1) $M + N_2 \rightarrow M_xN_y$

(2) $M_xN_y + H_2 \rightarrow M_xN_{y}\text{-}z + zNH_3$

Microwave & Thermal Catalytic Method

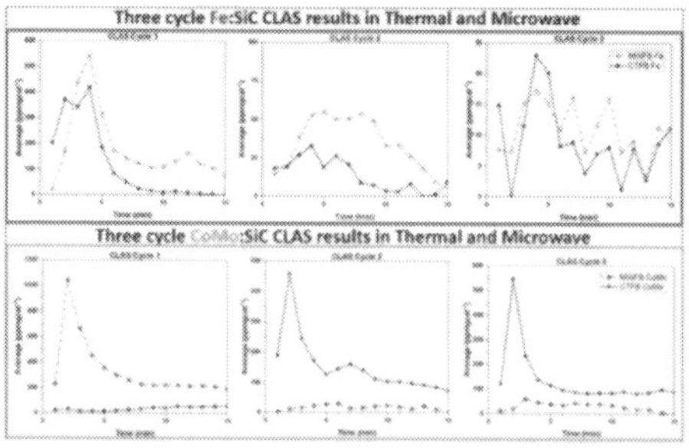

Plasma + Thermal Catalytic Method

- Pre-activation of N≡N, lower T
- Surface modification + promotion

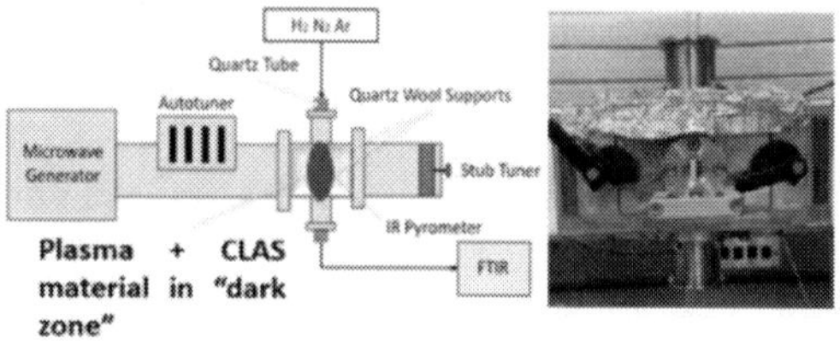

220c Pathways for NH_3 Formation from N, H, and $N_2(v)$ in Plasma Catalysis

Brian Bayer, Peter Bruggeman, Aditya Bhan

NH_3 synthesis by plasma catalysis at atmospheric pressure has been previously demonstrated [1], but pathways for NH_3 formation involving plasma-derived radical (N, H) and vibrationally excited species ($N_2(v)$) over catalytic surfaces have not been clearly identified. We measure densities of N, H, and $N_2(v)$ produced by an atmospheric pressure plasma jet using threshold ionization molecular beam mass spectrometric techniques [2-3] and correlate consumption of these species with NH_3 formation in a packed bed of nonporous metal wools of Fe, Ni, and Ag. Results show that formation of NH_3 and consumption of N, H, and $N_2(v)$ is faster over the catalytic surfaces than in the gas phase. N consumption correlates with NH_3 formation when the ratio between the H density and N density at the inlet is greater than 3 (stoichiometric for NH_3 formation) for all three metals. For Ni and Ag, N consumption is ~100% selective to NH_3 formation. For Fe, selectivity to NH_3 formation is dependent on the H_2 density. When the ratio between H density and N density at the inlet is less than 1, N consumption does not correlate with NH_3 formation for all three metals, indicating that NH_3 formation is limited by H density at these conditions. $N_2(v)$ consumption exhibits no correlation with NH_3 formation for any of the examined metals, indicating that surface-mediated vibrational relaxation of $N_2(v)$ is more significant than dissociative adsorption at these conditions. These correlations between NH_3 formation and consumption of plasma-derived intermediates, preliminary spectroscopic studies probing identity of catalyst-bound surface species formed in N_2/H_2 plasma, and design and operation of a reactor that enables these measurements will be discussed.

References

[1] M. L. Carreon, *J. Phys. D: Appl. Phys.* 52 (2019) 483001

[2] J. Jiang *et al.*, *J. Phys. D: Appl. Phys.* 55 (2022) 225206

[3] J. Jiang *et al.*, in preparation.

220d Accessing Meta-Stable States through Plasma Catalysis

Casey O'Brien, Garam Lee, David Go

Low temperature plasmas (LTPs) are partially ionized gases in which high energy electrons collide with gas molecules to produce a highly reactive chemical environment, including free electrons, ions, radicals, and vibrationally- and electronically-excited molecules, at bulk temperatures (~300-600 K) at which such species are thermally inaccessible. The integration of atmospheric pressure LTPs into catalytic systems has recently gained considerable attention because of its potential to carry out transformations that are difficult or impossible using conventional thermal catalysis (e.g. CH_4, N_2, and CO_2 activation), and in modular units powered by renewable electricity. In this talk, I will demonstrate using a newly developed multi-modal operando polarization-modulation infrared reflection absorption spectroscopy tool that plasma-enhanced catalysis can access meta-stable surface-adsorbed species that can only be accessed through the integration of plasma with catalytic surfaces. Specifically, we will demonstrate that, in the presence of nitrogen plasma, azide (N_3) species are stabilized on metal (e.g. Ni, Ag, Au) surfaces with extraordinary stability and unique reactivity. We will demonstrate that surface-adsorbed atomic nitrogen species, which are deposited onto the metal surfaces during exposure to N_2 plasmas, are critical for the formation of azide species. However, deposition of atomic nitrogen species is not alone sufficient to access the surface-adsorbed azide species. We will show that plasma-activated N_2 interacts with surface-adsorbed atomic nitrogen to form the corresponding azide species, which is stable up to ~500 K. Finally, we will demonstrate that this surface-adsorbed azide species has unique reactivity compared to atomic nitrogen. This work demonstrates that meta-stable states can be accessed by integrating plasmas with catalysts, and that these meta-stable states have unique reactivity that is not typically considered in plasma-enhanced reactions involving N_2.

220e Methane Dehydroaromatization over Mo/H-ZSM-5 Under Non-Thermal Plasma Stimulation

Gerardo Rivera-Castro, Jason Hicks

Non-thermal plasmas are promising alternatives to thermal processes that can activate stable gas molecules close to ambient temperatures by electron impact reactions. When a non-thermal plasma is coupled with a catalyst, reaction rates and selectivity towards desired products are further improved. Plasma activation approaches have been applied to the direct conversion of methane to value-added products since this stable molecule requires high energy input and elevated temperatures for activation. Methane dehydroaromatization (MDA) is a non-oxidative approach to directly convert methane into aromatics over a Mo/H-ZSM-5 catalyst, and an equilibrium conversion of 14% can be obtained at 973 K. However, high temperatures, coke deactivation and low-single pass methane conversions are major drawbacks associated with this process.

Here, a non-thermal plasma-catalytic methane dehydroaromatization process is presented, where we show that this chemistry can be performed at lower bulk gas temperatures, compared to the high temperatures required for methane conversion under thermal catalytic conditions. We observed that the bulk gas temperature directly influences the behavior of the plasma discharge, and led to lower methane conversions when catalysts were absent. However, incorporation of Mo/H-ZSM-5 in the plasma discharge significantly enhances methane conversion under these conditions, while shifting the product distribution toward the formation of aromatics. This presentation will highlight our current understanding of this plasma-catalytic methane dehydroaromatization mechanism and how plasma stimulation can facilitate methane dehydroaromatization pathways at lower reaction temperatures over a Mo/H-ZSM-5 catalyst.

220f Recombination after the Plasma and Product Distribution

Colin Page, Elijah Thimsen

Plasma has proven to be an effective medium to promote highly endothermic chemical reactions with reasonable energy efficiency. Two notable examples are carbon dioxide splitting into carbon monoxide and oxygen, and nonoxidative methane reforming into acetylene and hydrogen. These reactions can be carried out with energy efficiencies, based on the enthalpy of reaction, in the range 20 to 50%, at high volumetric productivity. Understanding of the key plasma process parameters that are required to promote exothermic reactions, by contrast, is much less developed. In this presentation, the focus will be on the background temperature in the recombination zone following the plasma. The hypothesis is that a lower background temperature after the plasma promotes the formation of products from relatively exothermic reactions and suppresses products from relatively endothermic reactions. Results will be presented from an experimental test of this hypothesis using nonoxidative methane reforming as an example system.

220g A Novel Pulsed-Plasma Catalytic Reactor for Dry Reforming of Methane

Michael Mullins, Ben Caithamer

Concerns over greenhouse gases have increased interest in the Dry Reforming of Methane (DMR) which produces hydrogen and carbon monoxide from the reaction of two greenhouse gases (CO_2 and CH_4) over a catalyst. Currently, DMR is primarily a catalytic process which operates at temperatures between 700°C - 900°C, and 10 to 20 bar using a 1–1.5 ratio of CH_4/CO_2. Unfortunately, these conditions also promote the water-gas shift reaction, which produces additional CO_2. We have developed a non-thermal, pulsed-plasma catalytic DRM reactor which operates at ambient temperatures and pressures. When combined with an integral monolithic catalyst bed this reactor demonstrated high conversions (60 to 95%) of both methane and carbon dioxide with high yields of hydrogen and carbon monoxide (50 to 80%). To achieve this, we developed a solid-state, MOSFET-based HV pulse generator with controllable rise times (<5 ns), pulse duration (0.1 to 10 ms), pulse shape, and frequency (10-10,000 Hz), providing improved operational flexibility and better energy efficiency than older pulse generating circuits. The reactor incorporates a point-to-plane electrode arrangement with an integral monolithic reactor which effectively couples the excited state plasma to the catalyst. The catalysts employed are stable binary metal oxides tailored for low-temperature plasma DRM reactions. Bench scale reactor tests were conducted using a feed of methane and carbon dioxide diluted in argon. To evaluate the reaction kinetics, the partial pressure of the reactants and products were measured in real time via an on-line mass spectrometer, while the excited state species were simultaneously monitored using emission spectrometry. Tests were made with the plasma alone, and the plasma plus 4 different catalyst formulations. No significant reactions were observed for the plasma without a catalyst, or for the catalyst without a plasma. The catalytic reaction kinetics were measured for a range of input power, voltages, pulse length andamp; frequency, and electrode geometries. The feed ratio of CO_2 to CH_4 was found to be of great significance in the overall conversion and the yield of hydrogen and CO, with near stoichiometric reactant ratios proving best. The stoichiometric ratio of carbon monoxide to hydrogen in the products (from ~1:1 up to 6:1) was dependent on the ratio of the metal oxides employed in the catalyst, and can be correlated to the strength of carbon dioxide adsorption on the catalyst surface. Based on the kinetic studies and emission spectroscopy results, we propose a surface moderated reaction model which explains the high reactant conversions and product yields observed. Estimates of the energy efficiency of the bench-scale process, and a design for a larger scale reactor, indicate the potential of this novel reactor for practical applications.

418b Understanding the Nanostructure and Performance of Alumina Atomic Layer Deposited Films on a Layered Cathode Oxide Surface Using Molecular Dynamics Simulation

Julie A. Nguyen, Abigayle Becker, Krishan Kanhaiya, Alan Weimer, Hendrik Heinz

Atomic layer deposition (ALD) on layered cathode oxides like $LiCoO_2$ and $LiNi_{1-y-z}Mn_yCo_zO_2$ (NMC) has been shown to stabilize the surface and extend the cycling lifetime of these cathode active materials in lithium-ion batteries. While low cycle numbers of ALD have been experimentally shown to be beneficial to the lithium-ion cathode stability, the causes and mechanism for this stability are still elusive. We have utilized all-atom molecular dynamics simulations with accurate models for lithium Ni-Mn-Co oxides, as part of the INTERFACE force field (IFF) to study the alumina film morphology and the distribution of surface lithium through the film. We derive correlations between the number of alumina layers, the film crystallinity, the film thermodynamic stability, the lithium availability for charge transport at the film-cathode oxide interface, and the experimentally observed battery performance in terms of discharge capacity and number of ALD cycles .

Specifically, we studied alumina (Al_2O_3) deposition in contact with a layered cathode oxide (e.g., $LiCoO_2$ or other NMCs) without specific chemical modifications (such as hydration or carbonation products), and we modeled the trimethylaluminum (TMA)/water ALD chemistry with the cathode oxide surface after complete reaction. IFF in the 12-6 Lennard Jones form was used to model both the alumina and the layered cathode oxide. Using molecular dynamics simulations, we studied the structure, energy, and lithium distribution for zero to six monolayers of Al_2O_3: including multiple initial structures and annealing protocols to sample the configuration space and various degrees of simulated X-Ray crystallinity of the ALD films.

We find alumina films containing two to three monolayers exhibit favorable binding energies to the layered cathode oxide surface and optimum lithium availability on the exterior of the alumina film. Alumina layer numbers outside of two to three layers are thermodynamically less favorable. Generally, the more thermodynamically stable thin film-cathode interfaces possess lower alumina film crystallinities with the exception of two layers, which is highly crystalline, while thicker films are more crystalline. The lithium density profiles through the films show that the lithium exists on either side of thin Al_2O_3 films while the lithium becomes increasingly incorporated in the bulk of thicker films.

These atomic scale findings help explain experimental data showing two to four alumina ALD cycles yield a maximum discharge capacity and longer cycling stability compared to the uncoated NMC oxide surfaces.

418c First-Principles Modeling of Discharge Product Surface Thermodynamics in Na-O_2 Batteries

Alex Von Gunten, Kunal Velinkar, Eranda Nikolla, Jeffrey Greeley

With the current climate crisis, the growing need for utilization of renewable energy resources necessitates the development of improved electrochemical storage devices. One technology that shows promise in this regard is aprotic alkali metal-O_2 batteries (Li/Na-O_2), which have greater than threefold higher theoretical gravimetric energy densities than state-of-the-art Li-ion batteries[1]. With a greater elemental abundance of sodium and low observed overpotential losses, Na-O_2 batteries have become a focus of research interest[2]. Despite recent efforts, however, poor cell lifespans continue to plague Na-O_2 cells (≈10s of cycles), remaining well below commercial viability[3]. A contributing factor to these poor cell lifespans is an inherent instability (parasitic reactivity) at the cathode between the desired discharge product (NaO_2) and the organic electrolyte, leading to undesirable side product formation and a rapid decline in cell capacity. The formation mechanism for these side products is unknown, and one hypothesis is that the surface of the discharge product first undergoes an undesirable dissolution into the electrolyte before subsequently forming the side products[4]. Thus far, practical solutions to improve stability have remained elusive. By elucidating the molecular-level mechanisms of these deleterious processes, using both theory-driven and experimental approaches, it may be possible to overcome these bottlenecks.

The goal of the present work is to couple computational and experimental studies to elucidate key molecular-level mechanistic insights of the parasitic reactivity occurring at the surface of the NaO_2 discharge product. We utilize Density Functional Theory (DFT) calculations to study the surface thermodynamics of the NaO_2 discharge product, of which the knowledge significantly trails that of the closely related Li-O_2 battery chemistry. We elucidate thermodynamically relevant surface facets and surface layer stoichiometries by constructing an ab-initio surface phase diagram of the NaO_2 discharge product. We discover an inverse relationship between coordination and stability of surface terminations under vacuum. We further observe a competing energetic effect between smoothness and packing density at the surface layer. In addition, we identify a representative set of surface features (including terraces, steps, and defects) for use in subsequent studies of parasitic reactivity mechanisms at the NaO_2 discharge product surface.

As an extension of the surface phase diagram, an ab-initio mechanistic surface dissolution analysis is applied to the identified surface terminations, probing the effect of coordination on the thermodynamics governing surface dissolution. In this analysis, the thermodynamic energy barriers are compared between surface terminations, and as with the surface stability, we discover that there is an inverse relationship between coordination and the thermodynamic barrier for surface dissolution. In considering both stoichiometric and off-stoichiometric pathways, we find dissolution will follow a concerted pathway, such that stoichiometric NaO_2 units will dissolute from the surface, rather than as individual Na^+ and O_2^- ions. In closing, we comment how the current work lays the groundwork for future combined computational-experimental studies and how they may be used to propose practical design changes.

References:

(1) Samira, S.; Deshpande, S.; Greeley, J.; Nikolla, E. Aprotic Alkali Metal–O_2 Batteries: Role of Cathode Surface-Mediated Processes and Heterogeneous Electrocatalysis. *ACS Energy Letters* **2021**, *6* (2). https://doi.org/10.1021/acsenergylett.0c02506.

(2) Kwak, W.-J.; Rosy; Sharon, D.; Xia, C.; Kim, H.; Johnson, L. R.; Bruce, P. G.; Nazar, L. F.; Sun, Y.-K.; Frimer, A. A.; Noked, M.; Freunberger, S. A.; Aurbach, D. Lithium–Oxygen Batteries and Related Systems: Potential, Status, and Future. *Chemical Reviews* **2020**, *120* (14). https://doi.org/10.1021/acs.chemrev.9b00609.

(3) Sun, Q.; Liu, J.; Xiao, B.; Wang, B.; Banis, M.; Yadegari, H.; Adair, K. R.; Li, R.; Sun, X. Visualizing the Oxidation Mechanism and Morphological Evolution of the Cubic-Shaped Superoxide Discharge Product in Na–Air Batteries. *Advanced Functional Materials* **2019**, *29* (13). https://doi.org/10.1002/adfm.201808332.

(4) Kim, J.; Park, H.; Lee, B.; Seong, W. M.; Lim, H.-D.; Bae, Y.; Kim, H.; Kim, W. K.; Ryu, K. H.; Kang, K. Dissolution and Ionization of Sodium Superoxide in Sodium–Oxygen Batteries. *Nature Communications* **2016**, *7* (1). https://doi.org/10.1038/ncomms10670.

418d Interfacial Stability of Al and Ga Dopants at the Lithium Garnet | Lithium Metal Interface

Matthew Klenk, Michael Counihan, Zachary Hood, Justin Connell, Sanja Tepavcevic, Peter Zapol

All-solid-state lithium-ion batteries (ASSB) are a critical technology for the future development and expansion of lightweight, energy dense, and safer battery technologies. The move to an ASSB would allow for the integration of higher energy density electrode components in particular Li metal as an anode. Of the potential solid electrolyte materials, the lithium garnet, $Li_7La_3Zr_2O_{12}$ (LLZO) has been touted as a leading candidate due to its high ionic conductivity (10^{-3} S/cm), toughness (shear moduli ~60GPa) and purportedly strong electrical and chemical stability (0.02-4.0 V vs Li/Li^+). However, progress in advancing the deployment of ASSB has been slow due to the difficulty of engineering a stable solid-solid interface between the solid electrolyte and lithium metal. This is further complicated in the garnet system where a wide range of dopants are commonly used further expanding the possibility of impurity phase formation at these interfaces.

Previous works looking at the stability of dopants on the Zr site (Zr, Ta, Nb), observed that reduction of these transition metals is possible the presence of Li metal. This can result in the formation of impurity phases as the reduction of dopant species is propagated through the microstructure. Luckily, methods to control lithium stoichiometry, and stabilize cubic LLZO, are not limited to Zr site doping. Strong electrochemical performance has been obtained by doping the Li_{24d} site via Al and Ga substitution. These alternatives to Ta or Nb allow for the removal of three Li ions per Al/Ga ion insertion and are commonly believed to be chemically equivalent to each other. Interestingly, the ionic conductivity for gallium doped samples is reported to be 2x-5x greater than aluminum doped samples. One possible explanation for this is the formation of Li-Ga eutectics that enhance the interfacial contact between the Li metal and solid electrolyte. In this study we look to investigate the stability of aluminum and gallium dopants at the $LLZO|Li^0$ interface through a combined effort of theoretical modeling and experimental techniques including, XPS, impedance spectroscopy, neutron diffraction.

During the structural optimization of Al/Ga-LLZO|Li interface simulation cells, regardless of the distance from the interface, Al doped structures optimized to relatively the same energy. However, using the same initial atomic distributions and replacing Al with Ga, results in two regions of energy convergence. The Ga that are located at the Li metal interface or on the free surface have converged to energies lower than configurations where Ga is in the "bulk" of LLZO. Additional simulations where Ga is imbedded in the Li metal layer, result in even lower energies, suggesting a stable Li-Ga alloy formation at the interface. The difference surface/interface stability between the two dopants appears to be related to their relative bonding affinity in oxides verses native metal or alloy compositions.

When in contact with E-beam deposited Li metal, XPS measurements on Ga and Al doped LLZO samples show that there are additional peaks characteristic to Ga reduction, while no additional peaks are evident in Al doped samples. Unlike the observed reduction in Zr site species, the reduction of Ga occurs under low energy E-beam deposition, indicating that the Ga is easily reduced in this system. The reactivity of Ga dopants in LLZO is further supplemented by our impedance spectroscopy work looking at the aging of Ga doped LLZO samples in contact with Li metal. During the aging process we can divide the observed degradation of the cell into three regimes. Region one where the initial Li-Ga allow is formed resulting in lowered interfacial resistance. Region two, when the reduction of Ga is slowed, and there is an increase in the charge transfer resistance. Finally, once all easily accessible Ga is depleted, region three is entered where further degradation of the cell has halted with the formation of a equilibrium interface.

The observed instability of the Ga-LLZO compared to Al-LLZO is critical to progressing our understanding of engineered interfaces in solid-state batteries. Despite the improved ionic conductivity of Ga doped samples, the chemical instability of Ga makes it a poor choice compared to Al, which forms a more stable interfacial structure. Further, techniques that look to form interfacial Li-Ga alloys through liquid metal processing at the Li|LLZO interface may be in vain, where the interphase compositions are similar to the naturally formed interface structures observed in this work. The formation of Li dendrites in the microstructure during the cycling of LLZO batteries may be directly related to the preference of Ga at particle surfaces. Because Ga is more stable at the surface of a particle, it is more likely to be in contact with lithium metal, increasing the probability that the dopant to be reduced. As Ga is reduced and incorporated in a Li-Ga alloy, it is uncertain whether Li is reintercalated into LLZO or precipitated out to the interface.

418f Synergistic Effects between Geometric and Electronic Factors on Crystallized Polymer-Coated Pt Catalyst for Hydrogen Evolution Reaction

Chi-Ho Lee, Jung-Hyun Park, Hong Yang, Joseph Kwon

One way to deal with the growing energy demands and the vulnerability to sustainable energy is to discover the energy resources that are environmentally friendly, renewable, and plentiful.[1] Among several materials with the above characteristics, hydrogen has been considered a key fuel in the design of innovative energy generation and storage technologies.[2-3] One very widely adopted way to produce hydrogen is a half-reaction of electrochemical water splitting called hydrogen evolution reaction (HER) which requires an appropriate catalyst to enhance the reaction efficiency.[4-5] In this reaction, platinum (Pt) has been widely used due to its ideal *d*-orbital tendency leading to a low overpotential value.

However, although Pt-based materials have an excellent electronic feature with a metallic nature, there has been a need for improvement of the desorption rate (Tafel reaction) of hydrogen gas compared to adsorption rate (Volmer reaction) of hydrogen. For the above-mentioned reason, we newly designed diverse crystallized polymer-coated Pt catalysts to significantly enhance the Tafel reaction for hydrogen gas desorption, and suggested a candidate material that surpasses the conventional Pt. Further, to obtain theoretical insights into the catalytic activity priority of the designed structures, we systematically analyzed the correlations between hydrogen binding strength (ΔG_{H*}) and electronic/geometric properties such as electron charge state, *p*-/*d*-orbital variation, electrostatic interaction, and strain effect. These theoretical investigations were able to clearly describe the origin of enhanced HER performance by boosting Tafel reaction. In addition, our results are consistent with the experimental activities of the newly designed structures and their activity order from thermodynamic and kinetic viewpoints.

References

[1] W. Lubitz and W. Tumas, *Chem Rev*, (107), 2007, pp. 3900-3903.

[2] N. L. Panwar, S. C. Kaushik and S. Kothari, *Renew Sust Energ Rev*, (15), 2011, pp. 1513-1524.

[3] Z. G. Yang, J. L. Zhang, M. C. W. Kintner-Meyer, X. C. Lu, D. W. Choi, J. P. Lemmon and J. Liu, *Chem Rev*, (111), 2011, pp. 3577-3613.

[4] M. S. Dresselhaus and I. L. Thomas, *Nature*, (414), 2001, pp. 332-337.

[5] Y. Jiao, Y. Zheng, M. T. Jaroniec and S. Z. Qiao, *Chem Soc Rev*, (44), 2015, pp. 2060-2086

418g Insights into the Effects of Micropore Structure and Electrolyte Composition on the Energy Storage Performance of Electrical Double Layer Capacitors

Betul Uralcan

Electrical double layer capacitors (EDLCs) store energy in the form of electrical charges at the interface between an electrolyte and a high surface area electrode. Despite having high power density and long cycle life, widespread adoption of EDLCs has been restricted due to their limited energy density. Optimizing EDLCs energy storage performance is thus desirable to promote their competitiveness in energy storage applications. Here, we conduct constant potential molecular dynamics simulations of realistically modeled nanoporous carbon electrodes and ionic solutions to explore the effect of electrode micropore structure and solvent concentration on the energy storage kinetics and equilibrium properties of EDLCs. We show that when average pore size is similar to or slightly larger than the size of a solvated ion, solvation weakens anion-cation coordination in nanopores and enhances ion electrosorption by lowering the free energy barrier for counterion penetration while blocking the coions. Areal capacitance of these systems shows a nonmonotonic dependence to solvent concentration, exhibiting an initial increase with solvent introduction in the concentrated regime followed by a decrease with further dilution. This gives rise to a maximum in capacitance at intermediate dilution levels. When pores are significantly larger than solvated ions, capacitance maximum weakens and eventually disappears. Our results also demonstrate that charging profiles of electrodes with larger pores reach the plateau regime faster while charging time has a nonmonotonic dependence to ion concentration that mirrors the composition dependence of bulk electrolyte conductivity. These findings provide novel insights on the charging kinetics and equilibrium behavior of realistically modeled electrical double layer capacitors. Generalization of the approach presented here can facilitate rational optimization of material properties for electrical double layer capacitors.

Molecular insights on the performance of nanoporous carbide-derived carbon supercapacitors with various electrolytes

Xiaobo Lin[1,2], Shern R. Tee[3], Peter T. Cummings[1,2]
[1]*Multiscale Modeling and Simulation Center, Vanderbilt University, Nashville, TN, USA*
[2]*Department of Chemical and Biomolecular Engineering, Vanderbilt University, Nashville, TN, USA*
[3]*Australian Institute for Bioengineering and Nanotechnology, The University of Queensland, Brisbane, Queensland, Australia*

ABSTRACT

Supercapacitors, also known as electric double layer capacitors (EDLCs), have drawn increasing interest due to their rapid charging capability and long cycle lifetimes. This charge storage mechanism has no solid phase transport limitations, leading to high power density. However, EDLCs have lower energy density than batteries since the charge is confined to the electrode surface.[1] To improve the energy density of supercapacitors without sacrificing power density, it is desirable to better understand the charging behavior of supercapacitors and discover novel materials for electrolytes and electrodes.

Carbide-derived carbon (CDC) electrodes are high-surface-area nanoporous materials, and ion desolvation happens in CDC pores smaller than the solvated ions, which dramatically improves the capacitance of CDC supercapacitors.[2] Fundamental research has been done to uncover the molecular phenomena and mechanism[3], but molecular studies for different electrolytes in CDC supercapacitors have not been reported. We use all-atom models to describe each molecule in the system, as opposed to coarse-grained models, which allows us to study different electrolytes and capture explicit atom-atom interaction information. We consider several promising electrolytes. For example, ionic liquid/organic solvent mixtures exhibit higher ion conductivity while retaining relatively wide electrochemical voltage stability windows. In addition, solvent-in-salt (SIS) electrolytes (i.e., highly concentrated salt solutions) are promising for next-generation high-energy density supercapacitors and batteries due to their expanded voltage windows, nonflammability, and lack of volatility.[4] We use constant potential molecular dynamics to investigate the effects of these electrolytes on charging dynamics and charge storage of realistically modeled CDC supercapacitors. This yields molecular insights into the charging dynamics and charge storage mechanism of supercapacitors with different electrolytes, allowing us to select novel electrolytes or electrodes to improve supercapacitor performance.

References

(1) Simon, P.; Gogotsi, Y.; Dunn, B. Where Do Batteries End and Supercapacitors Begin? *Science (80-.).* **2014,** *343* (March), 1210–1211.

(2) Simon, P.; Gogotsi, Y. Materials for Electrochemical Capacitors. *Mater. Sustain. Energy A Collect. Peer-Reviewed Res. Rev. Artic. from Nat. Publ. Gr.* **2010,** 138–147.

(3) Feng, G.; Cummings, P. T. Supercapacitor Capacitance Exhibits Oscillatory Behavior as a Function of Nanopore Size. *J. Phys. Chem. Lett.* **2011,** *2* (22), 2859–2864.

(4) Yamada, Y.; Furukawa, K.; Sodeyama, K.; Kikuchi, K.; Yaegashi, M.; Tateyama, Y.; Yamada, A. Unusual Stability of Acetonitrile-Based Superconcentrated Electrolytes for Fast-Charging Lithium-Ion Batteries. *J. Am. Chem. Soc.* **2014,** *136* (13), 5039–5046.

349b Advancing Quantitative Understanding of Functionality in the Lithium SEI in Liquid Electrolytes

Betar Gallant

Li metal anodes offer significantly higher capacities than graphite and are therefore central to strategies to develop advanced rechargeable battery chemistries that meet range and performance targets for electric vehicles. Although closer than ever, lithium (Li) anodes still cannot meet the >99.9% Coulombic efficiency (CE) consistently needed for >1,000 cycle life. This shortfall arises from uncontrolled reactivity at the solid electrolyte interphase (SEI) and its resulting properties, leading to inhomogeneous plating and stripping, continuous electrolyte consumption and loss of active Li inventory. Despite much recent progress in electrolyte development, the lack of quantitative understanding of functionality from the perspective of the SEI itself still hinders attempts to rationally design an improved interface and bridge the remaining gap in CE.

To help inform such efforts, our work is developing techniques to gain insights into SEI phases and reveal interplays between their chemistry, structure and function. First, we developed approaches to isolate and synthesize SEI-relevant ionic phases, including lithium oxide (Li_2O), lithium fluoride (LiF) and others, at representative nanometer-scale thicknesses directly on Li metal. These interfaces are interrogated *via* targeted electrochemical and spectroscopy techniques to reveal their transport properties, Li^+ exchange kinetics and chemical reactivity, providing insight into how such phases bolster, or hinder, transport in the SEI. Recently, quantification efforts have been extended towards characterization of the native SEI. Using a combination of voltammetry and electrochemical impedance spectroscopy-based methods, we report on self-consistent Li^+ exchange rates across a comprehensive range of electrolytes and show that Li CE and rate capability are tightly correlated with the rate of SEI Li^+ exchange. The magnitudes of such exchange rates delineate critical electrolyte-dependent cycling rates beyond which deposition is interface-limited and CE plummets. For many electrolytes, this transition occurs around typical (~1 mA/cm^2) imposed cycling currents, elucidating the importance of cycling protocol and rate on Li reversibility. Moreover, we examine the highly electrolyte-dependent changes in Li^+ exchange upon Cu formation and cycling, which exhibit unexpected dynamics. These results are collectively helping to increase quantitative understanding of the SEI, which guides electrolyte and additive selection as well as design of cycling protocols to maximize CE.

349c On-Line Flow Cell Trace Elemental Analysis Elucidates Degradation Mechanisms in Transition Metal Oxygen Electroreduction Catalysts

Gaurav A. Kamat, Sanzeeda Shuchi, Melissa Kreider, Jose A. Zamora Zeledon, Michaela Burke Stevens, Thomas Jaramillo

Commercial proton exchange membrane hydrogen fuel cell cathodes are presently composed of platinum-based materials that exhibit exceptional oxygen reduction reaction (ORR) activity and stability. However, platinum electrodes remain prohibitively expensive for transitioning to a more sustainable economy dependent on hydrogen fuel cells. Less expensive transition metal ORR catalysts have been thoroughly investigated for identifying high activity alternatives with established methods though the stability of alternative materials is not as well-understood. Methods for assessing catalyst degradation typically involve surface characterization, electrolyte aliquot collection for metal quantification, and activity loss measurements, all performed after catalysis for extended periods of time. Experiments of these types can provide information about how catalyst materials degrade over time but usually cannot resolve mechanistic information or phenomena that occur on shorter time scales. More intensive experiments using synchrotron radiation have been utilized to probe degradation mechanisms in platinum-based fuel cell catalysts but investigations on alternative materials are limited due to the accessibility of synchrotron facilities and high time requirements for such experiments. In our work, we utilize an in-house technique consisting of an on-line electrochemical flow cell linked to an inductively coupled plasma-mass spectrometer (ICP-MS) for evaluation of fuel cell catalyst degradation in acidic media. Gas-saturated electrolytes of varying chemical identity and a constant pH of 1 flow through the cell and across a working electrode composed of platinum group-free metals. Dissolved metal ions are carried downstream to the ICP-MS where they can be measured with part per trillion resolution and minimal signal broadening, enabling millisecond resolution of degradation processes occurring at the catalyst surface. We evaluate the oxygen mass transport characteristics of a commercial flow cell against a custom-machined flow cell and optimize experimental conditions for comparability to the established rotating disk electrode (RDE) technique. Measurements on metal foils with applied cycled potentials in oxygen-saturated electrolyte display potential dependent corrosion similar to what we would expect from a typical Pourbaix (potential-pH) diagram of stability. However, cycling these metals to ORR-relevant potentials demonstrates degradation during ORR that increases with ORR current to varying degrees depending on the metal in a way that has not been previously characterized. Much of the degradation in the ORR potential range is not present when saturating the electrolyte with nitrogen, implying that the presence of oxygen can accelerate degradation. Probing the phenomenon of degradation during ORR further with differing scan rates and directions for the applied potential sweep reveals more information about the timescale of surface corrosion processes and informs the development of protocols that can enhance the lifetime of these catalyst materials under operation conditions. In combination with on-line testing in various pH 1 electrolytes, we make specific recommendations for optimal potential ranges and electrolyte compositions that optimize activity and stability to advise the fuel cell community in developing platinum group metal-free fuel cell cathode materials. We are also developing new flow cell designs to enable the testing of other non-precious ORR catalyst morphologies such as nanoparticle catalysts. The on-line ICP-MS flow cell platform demonstrates promise for accelerating materials stability analysis for next generation platinum-free ORR catalysts through practical experiments and direct comparability to established electrochemical techniques.

349d Exploring How Interfacial Ion Assembly Accelerates Electrochemical CO_2 Reduction

Matthew Gebbie, Beichen Liu, Wenxiao Guo

Electrochemical reduction of CO_2 to CO and other valuable products provides opportunities to realize sustainable production of fuels and chemicals. While demonstrated in lab-scale systems, a critical roadblock slowing development of industrial processes is poor understanding of how molecular assembly at electrode-electrolyte interfaces contributes to catalytic activity. Ionic liquids emerged as promising CO_2 reduction electrolytes, since low overpotentials and enhanced product selectivity has been observed for several classes of ionic liquids. However, mechanisms of ionic liquid-mediated catalysis are subject to ongoing discussion. Here, we present our research on tuning the collective assembly of ionic liquids at interfaces to understand ionic liquid-mediated CO_2 reduction at silver surfaces. By tuning ion structures, concentrations, and solvent environments, we reveal that CO_2 reduction to CO in ionic liquids can be accelerated by threefold over prior reports and find unexpected scaling relationships between structure, concentration, and reactivity. Further, we use *in situ* analysis of electric double layer formation at interfaces, including surface enhanced Raman spectroscopy and surface forces measurements, to yield mechanistic insights into how interfacial ion assembly can accelerate electron transfer to charged intermediates. Notably, our findings point to collective ion assembly as a key factor underpinning structure-reactivity relationships for CO_2 electrochemical reduction. To highlight the impact of this concept, we will discuss how our findings help resolve mechanistic puzzles in ionic liquids electrocatalysis and suggest new strategies for controlling ion and electron flow at interfaces to enhance the conversion of atmospheric CO_2 to CO and other fuel precursors.

349f Dynamic Photo-Electrochemical Interface for Solar Fuels

Weilai Yu, Nathan S. Lewis

Semiconductor photoelectrochemistry (PEC) is a promising route for sustainable production of clean H_2 fuel vis solar water splitting However, scalable implementation of high-efficiency PEC device is plagued by their limited stability. For decades, stability has remained one of the most significant but least understood topics in the field. Herein, I will introduce our efforts at Caltech to establish fundamental understanding over material corrosion and photoelectrode stability, for several technologically important semiconductors (InP, GaAs and GaInP) used as H_2-evolving photocathodes. Comprehensive insights are achieved by leveraging delicate experiment design as well as correlative measurements of electrochemistry, material dissolution as well as evolving surface conditions. Overall, three key concepts regarding stability are emphasized: (a) kinetic stabilization by surface catalysis, (b) Fermi level pinning by surface non-stoichiometry, and (c) structural integrity of multi-layer architectures in solar-fuel devices. Together, this advanced understanding reveals the dynamic nature of photo-electrochemical interface, whereas the electrochemical stability and physical stability of photoelectrodes need to be considered separately.

527a Advancing Nondestructive *in Situ* Infrared Nanospectroscopy of Buried Electrochemical Interfaces in Li-Ion and Li-Metal Solid State Batteries

Robert Kostecki, Hans Bechtel, Jonathan Larson, Xin He

The interface between solid/solid and solid/liquid phases plays a fundamental role in natural and engineered materials. In fact, most rechargeable batteries performance is largely determined by thermodynamic, kinetic, and mechanical properties of such electrochemical interfaces. Furthermore, secondary batteries operate far away from equilibrium and a thin passive film tend to form at the electrode/electrolyte interface. Situated at the surface of the electrode, and formed during battery assembly and/or operation, this so-called solid electrolyte interphase (SEI) layer is critical for the battery stable function and operation. The SEI's inhomogeneous structure and chemistry influence localized current density distribution and the resulting slow electrode and device degradation during charge–discharge processes. Our overall understanding of heterogeneous ionic interfaces and interphases is still very limited due to their chemical and physical complexity and lack of effective *in operando* characterization techniques. In this work, we exploit the nanoscale spatial resolution, chemical selectivity, and surface sensitivity of near-field infrared nano-spectroscopy, XPS and ATR FTIR to characterize buried interfaces in model carbonaceous Li-ion anodes and Li -metal electrodes. We demonstrate that even an atomically flat surface of graphene in contact with an SPE is still prone to develop a heterogeneous surface film composition and structure, which then results in nonuniform Li-plating currents and distributions at the nano- and micro-scale. This study provides a unique insight into the mechanisms of early-stage interphase formation at electrochemically active buried interfaces, and an experimental diagnostic means to aid in the development of methods to control local nanoscale variations in electrolyte chemistry, structure, and ionic conductivity at the surface of the electrode. Our approach also paves a way for interface optimization by providing a method for unprecedented nanoscale measurements of crystallinity, structure, conductivity, and chemistry of intact and buried electrochemical interfaces in their native environments.

527b Plasma-Liquid Interface for Promoting Organic Electrosynthesis without Solid Electrodes

Xiaoshuang Chen, Aditya Ponukumati, Marcus Foston, Elijah Thimsen

Plasma-liquid interfaces present an opportunity to promote electrosynthesis reactions in organic chemistry without the need for solid electrodes. The plasma-liquid interface may reduce deleterious effects of electrode fouling common in standard three-electrode configurations and may also avail reactants such as solvated electron without the need for hazardous reagents. To date, most plasma-liquid research has been conducted in aqueous media, which are of limited utility in organic chemistry. In this work, we focus on acetonitrile as the solvent, which has a wide electrochemical stability window and is useful in a wide range of organic chemical reaction systems. We demonstrate that the electrochemical structure of the plasma-liquid interface with acetonitrile solvent and a KPF_6 electrolyte is qualitatively similar to aqueous electrolytes. Furthermore, the solvent appears to be robust with respect to degradation over a range of plasma parameters. If the power supplied to the plasma was too high, then the solvent did start to decompose. Results from experiments conducted with a colorimetric redox indicator demonstrated that organic reduction reactions could be promoted by the plasma liquid interface, without the need for solid electrodes, similar to previous work done in aqueous systems.

527c Harnessing Spontaneous Electric Fields to Direct Thermochemical Catalysis

Thejas Wesley, Yuriy Roman, Yogesh Surendranath

While it is known that intrinsic electric fields play an important role in molecular and biological catalysis, their role in heterogeneous thermochemical systems remains unclear because the catalysts employed are typically disconnected from an external circuit (thus making it difficult to monitor or control the degree of electrical polarization of the surface). Here, we address this knowledge gap by developing general methods for wirelessly monitoring and controlling spontaneous electrical polarization at conductive catalysts dispersed in liquid media. By combining electrochemical and spectroscopic measurements, we demonstrate that proton and electron transfer from solution controllably, spontaneously, and wirelessly polarize Pt surfaces during thermochemical catalysis. Next, we rigorously show that the rates for ethylene hydrogenation using Pt-based catalysts are significantly influenced by spontaneous electric fields generated by both interfacial proton transfer in water and interfacial electron transfer from organometallic redox buffers aprotic solvents. We combine mechanistic experiments with transition state theory to elucidate the rigorous thermodynamic and molecular origin of this electrostatic phenomenology, and conclude that spontaneous polarization plays an essential role in non-faradaic reactivity at solid–liquid interfaces. This work shows that spontaneous interfacial electric fields should be considered alongside temperature, reactant activities, and catalyst structure as critical parameters to understand liquid-phase heterogeneous catalysis.

527d Saponin-Based Surfactant: Chemistry, Properties, and Applications on Enhanced OIL Recovery – Towards Sustainability

Lourdes Orejuela Escobar, Inés Hernández-Guerra, Juan Carlos Pereira, Yanet Villasana, Andrea Landázuri, Arleth Gualle Brito

Surfactants are one of the most versatile compounds due to their wide field of industrial applications. The utilization of synthetic surfactants is growing globally, being more than 15 million tonnes/year. However, around 60% ends in aquatic streams and marine environments, generating adverse environmental effects and causing contamination. Hence, it is imperative to seek more environmentally-friendly [1]. Hence, it is essential to seek novel natural sources as an alternative to synthetic ones but with similar surfactant properties [2]. Saponins are secondary compounds, heterosides, and phytochemicals formed by various aglycones or triterpenoid steroids and one or more carbohydrate moieties. Also called sapogenin and glucosaponins. Typically, they have high molecular weight and complex structures, giving them a variety of physical, chemical, and biological properties. Widely distributed in the plant kingdom, they are rich in functional compounds such as surfactants with potential and numerous applications such as moisturizing, foaming, detergent, emulsifying, solubilizing, and hemolytic properties, and others; among which the ability to form foam stands out. Historically, various compounds have been developed, based on saponins properties, having commercial potential in different industries such as food, chemical, cosmetic, pharmaceutical, textile, paints, lubricants, and cleaning products. However, various studies have established a more precise saponin definition based on its molecular structure and biosynthetic origin [3].

Similarly, research around the world has developed novel processes of extraction techniques, purification strategies, and technological applications [4]. The main objective of this study is to determine the interfacial behavior of a natural surfactant extracted from quinoa (*Chenopodium quinoa*) residues and determine its ability to produce foaming fluids within an oil reservoir to sweep large portions and recover larger quantities of crude oil from the reservoirs. Hence, this work aims to evaluate the efficiency of a natural saponin-based surfactant to minimize environmental impact and maximize cost efficiency as an ecologically sustainable replacement for synthetic surfactants.

The wet saponification was performed using the quinoa saponin extraction method, and spectrophotometric analysis was used for quantification. The quinoa residue samples were initially weighed, sieved, and mixed with distilled water using a magnetic stirrer, followed by vacuum filtration, centrifugation, and storage (Figure 1a). Subsequently, the methodology proposed by Monje, Yarko, and Raffaillac [5] was followed to quantify the percentage of quinoa saponins by applying the spectrophotometric method based on the construction of the calibration curve from the saponin sample Sigma-Aldrich 47036- 50G-F (standard) as a reference profile. Liebermann-Buchard (LB) color reagent was prepared and obtained by mixing acetic anhydride and concentrated sulfuric acid (1:5). Dilutions of the commercial sample were prepared at 0.1; 0.20; 0.3; 0.4; 0.5; 0.6 and 0.7 mg/mL; 0.600 mL of each were taken, to which 1 mL of LB the reagent was added, stirred for 20 s in a vortex, and left to stand for 37 min. Then, the calibration curve was built, measuring the absorbance for the different concentrations, at a wavelength of 528 nm, with a reading speed of 1 nm/s. An aliquot of the quinoa saponin samples was taken to be analyzed. Finally, the parameter R5* (hfoam t=5min-hfoam t=0min) was determined to estimate the stability of the foam.

Two analyses confirmed the chemical and physical characteristics of the saponin extract as a surfactant. FITR analyses were performed using Agilent Technologies Cay 630 equipment, following the KBr pillbox method, employing approximately 0.050 g of surfactant sample. The measurement parameters were: wavelength interval from 4000 to 600 cm^{-1}, resolution of 2 cm^{-1}, and eight scans. For the second analysis, the PerkinElmer STA 8000 equipment took 10 mg of the saponin sample. The test is heating started from 25 °C until reaching 800 °C at a rate of 5 °C/min, using 10 mL/min nitrogen flow as carrier gas. Subsequently, the flow of formulated foams was simulated and modeled for crude oil recovery by injecting it through a simple one-dimensional system in a heterogeneous porous medium. Several concentrations were initially prepared, varying salinity (NaCl), electrolyte charge (CaCl), pH (0.5 M citric acid or 0.5 M

NaOH), and quinoa saponin concentration (0.01; 0.1; 1 and 5%). Then, foams produced by commercial surfactants (Tween 80 and Tween 20) were also performed as a comparison.

The presence of saponins in the quinoa wash water was confirmed by physical observation by stirring, filtering, and rotary evaporating when foaming. In addition, the mixture of the quinoa washing water was tested with LB color reagent, presenting a pinkish-reddish color, characteristic of triterpenes (Figures 1b). The linear equation obtained by the calibration curve using the commercial saponin was: $Y = 6.2766 X - 0.1355$ and $R^2 = 0.9946$. This equation quantified the concentration of saponins in the analyzed samples, obtaining a range between 49% and 66.9%. However, the range reported in the literature varies from 47.3% to 56.2% [6]. Additionally, the relationship R5* parameter determined that the 0.5% saponin solution was 86% (metastable), representing excellent stability of the generated foam over time (Figures 1b).

The FTIR spectra of the saponins (Figures 1c) were similar to those of other already reported saponins [7]. Different signals or asymmetric stretching vibrations and the functional groups for this type of saponin were identified, with the alcoholic group at 3281 cm^{-1} and the active methyl group at 2942 cm^{-1}. However, between 2500 cm^{-1} and 1900 cm^{-1}, there is no absorption; that is, there is no accumulation of double and triple bonds, so the width of the band corresponds to the presence of hydrogen bonds, this peak confirms the presence of the C-O bond, and finally, the spectrum shows about 1377 cm^{-1} the presence of the C=C double bond, and at 1020 cm^{-1} a stretching signal C-O-C.

The TGA test showed the thermal stability of the saponins (Figures 1d). The pronounced mass loss occurred in three regions. The first stage ranged between 100 °C and 200 °C, attributable to the evaporation of moisture and surface water present in the natural surfactant, a range in which it presents a weight loss of 8.075%. In the second stage, the weight loss was 53.807%, which ranged between 100 °C and 500 °C, attributed to loss of chemically bound water, at which point carbohydrates are broken down. Finally, in the third and last stage, between 500 °C and 800 °C, the weight loss is accelerated by breaking the carbon bonds at high temperatures, with a weight loss of 9.566%. In summary, from the TGA test, it follows that the most significant weight loss occurs between 200 and 400 °C, so this implies that natural surfactant can be used in the oil field, where the reported reservoir temperature is between 60 and 150 °C since in this range the natural surfactant has adequate thermal stability.

The next step in this research will be to evaluate the influence of foam injection (formulated using saponin as a natural surfactant) in a porous medium as an improved hydrocarbon recovery technique "EOR" (Figures 1e), as an ecological, low-cost surfactant, and raw material alternative for sustainability production.

In conclusion, when mechanically shaken, the analyzed samples formed foam, evidencing the presence of triterpene type saponin (LB color reagent test). Saponin was quantified between 49 and 66.9%. It was found that the parameter R5* is over 50%, which indicates that the foams are stable over time. Different asymmetric stretching signals and functional groups for this type of saponin were identified through the FTIR test. The TGA test showed that the most significant weight loss occurs between 200 and 400°C, implying that natural surfactants can be used in oil fields, where the reservoir temperature ranges between 60 and 150°C, as an improved hydrocarbon recovery technique and as an alternative for sustainability production, ecological and low-cost technique.

References

[1] P. Johnson, V. J. Pinfield, V. Starov, and A. Trybala, "Effect of synthetic surfactants on the environment and the potential for substitution by biosurfactants," Adv. Colloid Interface Sci., p. 102340, 2020, doi: 10.1016/j.cis.2020.102340.

[2] D. Kregiel, J. Berlowska, I. Witonska, H. Antolak, C. Proestos, and M. Babic, "Saponin-Based , Biological-Active Surfactants from Plants."

[3] A. Troisi, J., Fiore, R., Pulvento. C., D´Andria, R., Vega-gález. A., Miranda, M., Martínez, E. y Lavini, "Saponinas. Capitulo Numero 3.3. IN: BAZILE D. et al. (Laboratorio Chimico Merceologico.," in Estado del arte de la quinua en el mundo en 2013: FAO (Santiago de Chile) y CIRAD, Montpellier, 2014.

[4] A. A. Muir, A.D., Paton,D.,Ballantyne, K., and Aubin, "Process for recovery and purification of saponins and sapogenins from quinoa," Chenopodium quinoa, 2002.

[5] R. J. P. Monje C. Y., "Determination of total saponin in quinoa (Chenopodium Quinoa Will) Spectrophotometric method," Mem. IV Natl. Congr. Boliv. Plant Prot. Assoc. Oru. April 5 to 7, vol. . C.E.A.C., 2006.

[6] M. Lozano, E. Ticona, C. Carrasco, Y. Flores, and G. R. Almanza, "CUANTIFICACIÓN DE SAPONINAS EN RESIDUOS DE QUINUA REAL," no. 2, pp. 131–138, 2012.

[7] M. S. Almutairi and M. Ali, "Natural Product Research : Formerly Natural Product Letters Direct detection of saponins in crude extracts of soapnuts by FTIR," no. March 2015, pp. 37–41, 2014, doi: 10.1080/14786419.2014.992345.

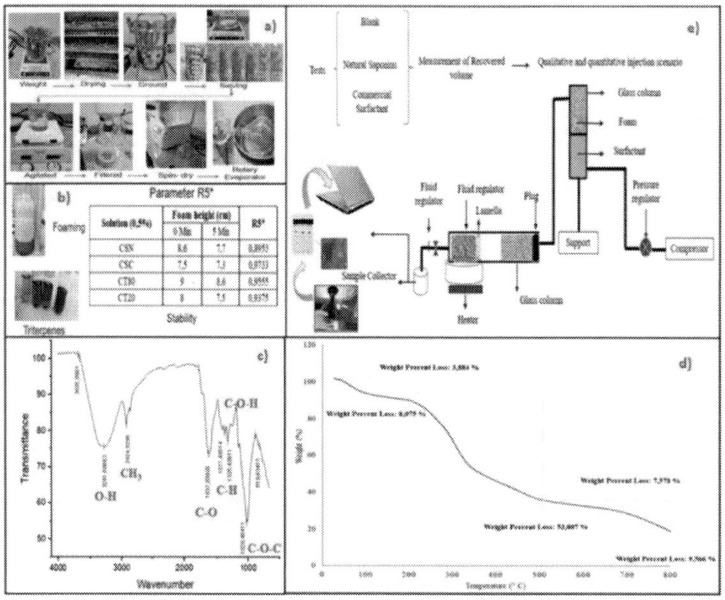

Figure 1. The Saponins: chemistry, properties, and applications on EOR

527e Determining Solvation Energies for Aqueous-Phase Reaction Adsorbates Using Ab-Initio Molecular Dynamic Simulations

Ankita Morankar, Hee-Joon Chun, Zhenhua Zeng, Jeffrey Greeley

The activity and selectivity of electrocatalytic reactions are governed by the interplay between the covalent reactant-surface bonding and additional interactions within the solid-liquid interface. Although the former interactions are well established in computational gas phase catalysis, the latter require further methodological development due to the complexity of electrochemical interfaces. In this study, we provide a simple theory for estimating the enthalpy of solvation for adsorbed intermediate(s). In a strategy reminiscent of Born-Haber cycles, solvation is broken down into two sub-processes: i) vacancy creation by water reorganization, and ii) subsequent interaction of the intermediate adsorbed in the vacancy with surrounding solvent molecules. The magnitude of the solvation energy is, in turn, determined as a mean value from a statistical sampling of hydrated adsorbate-catalyst configurations obtained from simulated annealing of ab initio molecular dynamic simulations.

The application of this method is illustrated in the context of oxygen reduction catalysis on platinum and Fe-N-C catalyst surfaces. Among other observations, we find that OH^* intermediates interact strongly with one another when clustering around hydrophilic FeN_4 centers located in hydrophobic graphene surfaces. Interestingly, this behavior causes the water molecules to solvate such clusters of hydrophilic intermediates to a lesser extent than they solvate isolated intermediates, since these coadsorbates form intramolecular hydrogen bonds within the cluster, and the solvent effect is correspondingly less. In contrast, coadsorbed OH^* and OOH^* intermediates adopt quite different structures on the surface of hydrophilic Pt(111) in aqueous solvents, as compared to adsorption in vacuum, due to a comparable order of interaction energy with solvent water molecules compared to the coadsorbates. Irrespective of the affinity of the catalyst for water, hydrophilic adsorbates exhibit significant stabilizing solvation free energies that directly impact oxygen reduction catalysis on both Pt and Fe-N-C catalysts. We close by briefly demonstrating how the solvation energy scheme may be extended to more complex C1 and C2 intermediates with both hydrophobic alkyl groups and hydrophilic hydroxy groups, of relevance to the methanol and ethanol electrooxidation reactions, on Pt(111) and Pt(211) surfaces. The alkyl moieties in the intermediate have a destabilizing effect owing to non-favorable hydrophobic interactions with solvent water molecules, while the hydrophilic hydroxy groups, have a stabilizing effect with favorable H-bonds forming with the solvent molecules.

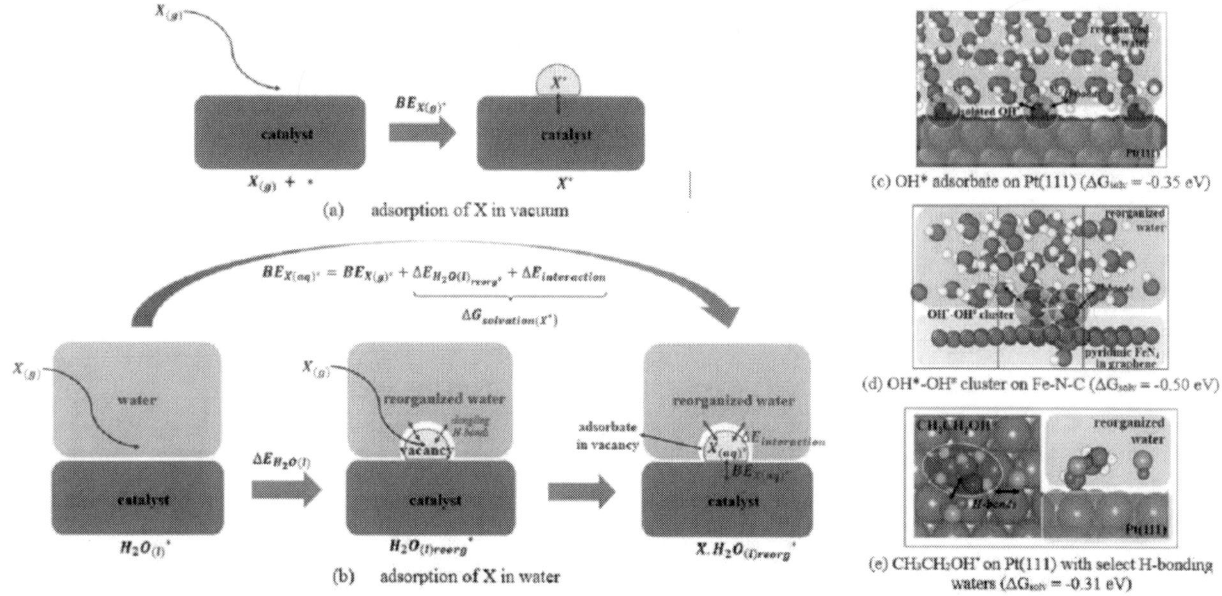

Figure 1: Schematics showing the adsorption process for: (a) X in vacuum, and (b) X in water with corresponding expressions for binding energies. Structures shown for (c) isolated OH^* on Pt(111), (d) OH^*-$2OH^\#$ cluster on Fe-N-C catalyst, (e) $CH_3CH_2OH^*$ on Pt(111)

574a Exploring Protein-Nanocrystal Interfaces for Photoassisted Enzymatic Activity

Jennifer Cha

With increasing demands for alternative sources of fuel, extensive research has focused on discovering methods to generate renewable energy from earth abundant resources. In recent years, a wide range of inorganic nanostructures with high surface areas and tunable band gaps have been synthesized and used as photocatalysts. In this talk, I will showcase our recent efforts to apply a bioinspired approach to tailor protein-nanoparticle interfaces for studying photoassisted redox activity. In order to drive enzymatic reduction of dissolved gases, electron sources such as photocatalysts and electrochemistry have been studied. However, in order to optimize electron flow, there is a significant need to understand how the interface between the organic enzyme and inorganic semiconductor influences protein binding and dynamics. Many redox enzymes function through assembly of protein subunits utilizing complex and multivalent interactions, with binding strengths ranging from long-range and weak to short-range and near-covalent. Mimicking such exquisite binding motifs is likely to be key for replacing protein subunits with photoactive semiconductors. This talk will showcase our recent insight into the design of interfaces between semiconductor surfaces and proteins to control binding and conformational dynamics of enzymes to promote photodriven redox catalytic activities.

574b Determining Rate Constants for the Nucleation and Growth of Silver Nanowires Via the Polyol Method in a Batch and Millifluidic Reactor

Destiny Williams, Shohreh Hemmati, James Smay

Silver nanowires (AgNWs) have proven to be promising candidates as the conductive material in conductive inks due to unique properties such as high conductivity, transparency, and mechanical flexibility. These conductive inks can be implemented in printing processes to create conductive patterns for use in manufacturing transparent conductive films (TCF). When considering a colloidal process for synthesizing AgNWs, the polyol method is a popular option because it is reliable, easy, and inexpensive. Polyol syntheses involve using a glycol as the solvent and the reducing agent, a silver metal precursor, a salt mediator, and a capping agent. AgNWs can easily be synthesized using the polyol method in a batch process, but these syntheses face difficulties when considered for industrial scale up. The most common problems faced are producing undesired silver nanoparticles (AgNPs), reproducibility, separation methods, and large amounts of waste. To address the issues associated with batch reactors, a millifluidic reactor can be used. Millifluidic reactors are continuous flow reactors constructed from tubing with inner diameters on the milli- scale and have a uniform chemical and thermal environment promoted by a fixed reaction volume inside the tubing. The kinetic study and determining the rate constants for the polyol synthesis of AgNWs is crucial for industrial scale up as they provide the ability to produce AgNWs with certain lengths and diameters. In this study, a method by El-Ghamry and Frei is implemented to quantify the amount of silver ions present at different times during the polyol AgNW synthesis in batch and millifluidic reactors and to compare the rate constants between the two types of reactors. Rate constants are determined by applying the non-linear Finke-Watzky model.

To apply the El-Ghamry and Frei method for quantifying silver ions, a calibration curve is made using an acetate buffer, solutions of eosin-y and 1,10-phenanthroline (TBF-Phen), and known concentrations of silver nitrate, the metal precursor. When the above components are combined for each known concentration of Ag, silver ions in solution will form Ag-TBF-Phen complexes that can be detected by UV-Vis with a characteristic peak at 550 nm. Once the calibration curve is established, polyol AgNW reactions are conducted in both batch and millifluidic reactors. For the polyol method, ethylene glycol (EG), silver nitrate, copper chloride, and polyvinylpyrrolidone (PVP) are used as the solvent/reducing agent, metal precursor, salt mediator, and capping agent, respectively. At a specific time-interval, a sample is taken from the reactor, diluted, and combined with the TBF-Phen solutions in a cuvette, and an absorbance value is recorded at the 550 nm peak. Silver ion concentrations are calculated using the calibration curve equation. Three sets of data are taken for each batch and millifluidic reaction, and the non-linear Finke-Watzky method for determining rate constants is applied. Using non-linear regression, rate constants are quantified and compared for each reactor.

This work is supported by the National Science Foundation (NSF) under grant number 1939018.

574d Synthesis and Stability of 2D Silicon Nanosheets

Matthew Panthani

Continued scientific and technological advancements in computing and microelectronics have resulted in massive improvements in healthcare, and technological advancement, and global development. However, the energy consumption that is attributed to information and communication technology is a growing concern. Integrated photonic circuits could dramatically reduce the energy cost of computing, while also improving computational speed. However, crystalline Si — the backbone of microelectronics—is a poor platform for integrated photonics due to its indirect bandgap, which makes light-emission a forbidden process. Existing optoelectronic materials are incompatible with our existing complementary metal-oxide-semiconductor infrastructure, which makes it difficult to fabricate integrated photonic circuits with the scalability and precision of Si-based microelectronics.

I will present my group's recent work in synthesizing nanostructured Si and other Group IV semiconductor nanomaterials, including nanocrystals and nanosheets with precise control over their size and surface chemistry. Using NMR, IR spectroscopy, and atomic pair distribution function analysis to gain insight in their local coordination and long-range order. The optoelectronic properties are sensitive to the structure and interfacial chemistry. Using ultrafast spectroscopy, there is evidence of a "direct-like" optical transition in 2D Si nanosheets, which is corroborated by DFT simulations of the band structure. I will also share recent work related to the thermal stability of 2D silicon nanosheets, demonstrating their structure and photoluminescence are stable to temperatures far beyond what they would experience in real applications.

References:

1. Ryan,B.J. et al. (2019) 'Silicene, siloxene, or silicane? Revealing the structure and optical properties of silicon nanosheets derived from calcium disilicide.' *Chemistry of Materials* 32(2) 795-804.
2. Ryan, B. J., Roling, L. T., andamp; Panthani, M. G. (2021). Anisotropic Disorder and Thermal Stability of Silicane. *ACS Nano*, *15*(9), 14557-14569.

154a Data-Driven Protein Design

Andrew Ferguson

Data-driven modeling and deep learning present powerful tools that are opening up new paradigms and opportunities in the understanding, discovery, and design of soft and biological materials. In this talk, I will describe our recent applications of deep representational learning to expose the sequence-function relationship within homologous protein families and to use these principles for the data-driven design and experimental testing of synthetic proteins with new and/or elevated function.

154b Low-Dimensional Fluctuating Interfaces for Efficient Separations

Narayana Aluru

Fluctuating, low-dimensional soft and hard material interfaces involve physical phenomena at various length scales. For example, the electronic structure of the low-dimensional material has a strong influence on the organization of the soft matter near the material. Atomic and molecular scale description of the materials is important to accurately describe the structure, dynamics, and thermodynamics of materials. Many nanomaterial applications involve macroscopic length scales that cannot be realistically modeled by atomic and molecular scale approaches. As a result, multiscale approaches linking various length scales are essential for many applications involving soft and hard nanomaterials. In this talk, we will present data-driven, machine learning-based multiscale approaches to model soft/hard nanomaterials. First, we will describe a deep learning-based multiscale approach to link ab-initio and molecular dynamics methods. We will show that the development of such methods can provide physical insights into nanomaterials that are not possible with purely ab-initio or molecular approaches. Second, we will describe deep learning-based multiscale approaches to coarse-grain materials at larger length scales. These methods can model systems at larger length scales and provide physical insights that are not possible with purely molecular approaches. Finally, using the data-driven multiscale approaches, we investigate dynamic interfaces involving low-dimensional materials and show that vibrational coupling has a significant effect on interfacial properties such as wetting, friction and fluid slip. We show how vibrational coupling can be tuned to design materials for efficient separations.

154c Enabling Catalyst Discovery through High-Throughput Experimentation and Machine Learning

Jochen Lauterbach

The development of high-throughput experimentation over the past two decades allows researchers to screen large compositional and operational spaces more efficiently and create large datasets under reproducible conditions [1]. Machine learning is an avenue to extract knowledge and unravel multidimensional relationships present in such large experimental datasets [2, 3]. A framework will be described that incorporates machine learning algorithms with experimental high-throughput catalytic data, spectroscopic data, and elemental properties to discover new materials. The framework allows to rank chemically descriptive features, to predict future catalyst performance, and to guide synthesis. Application of this framework predicted several novel catalyst compositions for low-temperature ammonia decomposition, which were experimentally validated against state-of-the-art ammonia decomposition catalysts and were found to have exceptional low-temperature performance at substantially lower weight loadings of Ru.

For many reactions, there is also a vast amount of scientific literature available over a large range of parameters. We extracted experimental data from the published literature for ammonia synthesis. However, gaps exist in the parameter space that have not been explored in the literature. These gaps can be rapidly filled using high-throughput measurements. Machine learning models were developed, trained, and compared to establish multi-dimensional correlations between catalyst formulations, elemental properties, support properties, synthesis parameters, reaction conditions, and ammonia synthesis rates. The activity of unknown catalyst formulations was predicted, and the best predicted catalyst candidates were synthesized and tested to validate the predictions. Consequently, this approach also led to the finding of new ammonia synthesis catalyst formulations with higher activity compared to some state-of-the-art catalysts in the literature.

1. Hattrick-Simpers, J.; Wen, C.; Lauterbach, J., The Materials Super Highway: Integrating High-Throughput Experimentation into Mapping the Catalysis Materials Genome. *Catalysis Letters* 2015, *145* (1), 290.
2. McCullough, K.; Williams, T.; Mingle, K.; Jamshidi, P.; Lauterbach, J., High-throughput experimentation meets artificial intelligence: a new pathway to catalyst discovery. *Phys Chem Chem Phys* 2020, *22* (20), 11174.
3. MacQueen, B.; Jayarathna, R.; Lauterbach, J., Knowledge extraction in catalysis utilizing design of experiments and machine learning. *Current Opinion in Chemical Engineering* 2022, *36*, 100781.

154d Predicting the Transition Temperature of Multi-Responsive Poly(N-isopropylacrylamide)-Based Microgels Using a Cluster-Based Partial Least Squares Modelling Approach

Seyed Saeid Tayebi, Prashant Mhaskar, Todd R. Hoare

A variety of applications in diverse fields has been made possible by smart microgels that can swell and deswell reversibly upon exposure to external stimuli, in particular temperature (by exploiting the volume phase transition temperature (VPTT) behaviour of microgels based on poly(N-isopropylacrylamide) or other thermoresponsive polymers) and/or pH (by incorporating ionizable functional groups into the microgel network). However, microgels with a targeted VPTT value are difficult to design because various physical and chemical factors affect both the VPTT and the ionization profile of microgels; dual pH/temperature-responsive microgels present even more complexities given that both stimuli contribute to the overall response of the microgel to either variable. To overcome these challenges, a data-driven model was designed and implemented in this work to predict the swelling profile and, ultimately, the VPTT of dual pH/temperature-responsive microgels based on copolymers of N-isopropylacrylamide and a variety of carboxylic acid-containing comonomers with different copolymerization kinetics and pK_a values. In the proposed clustering-based adaptation of Partial Least Squares (PLS) modelling, the data is grouped (either in the recipe space or the swelling profile space) and individual clusters are fitted to a PLS-based model. In the subsequent step, the method is applied to a data library of 33 microgels prepared with 8 recipe variables (i.e. the amount of co-monomers, crosslinkers, and surfactants) used as inputs to the model and 12 swelling profile variables measured at 6 temperatures at pH 4 (fully protonated regime) and pH 10 (fully ionized regime) as outputs to the model. To maximize the information extracted from the existing dataset, five different re-arrangements are also suggested under all three kinds of clustering policies (no clustering, recipe-based clustering, and swelling profile-based clustering). This approach could predict the microgels' transition temperatures within 0.8°C at pH=4 and 2.4°C at pH=10, an accuracy in line with the error bars associated with the measured transition temperatures (0.6°C and 2.2°C at pH=4 and pH=10, respectively). As such, the data-driven approach can accurately predict swelling transitions in dual-responsive microgels and, in tandem, predict microgel recipes that offers potential to achieve specific swelling/deswelling targets.

154e Designing Metamorphic Materials with a Machine-Learning Guided Genetic Algorithm

Kuan-Lin Chen, Rebecca Schulman

During natural selection, species evolve by generating large batches of variants and competing over numerous generations in search of optimized survival solutions. In the process, these systems learn the latent landscape and follow observed gradients toward optimization. The same idea can be largely adapted with machine learning algorithms and material discovery methods, designing automated systems that learn and optimize the design population.

Here, we demonstrate such a method for discovering designs for multi-stable metamorphic materials, DNA-responsive hydrogel architectures, that satisfy specific requirements for ready fabrication. Specifically, we ask whether it would be possible to fabricate metamorphic devices that, when presented with one of five or more different stimuli, can fold into five or more different functional shapes. We first develop a coarse-grained simulation that predicts the material conformations in response to different stimuli. We then train a machine learning model to score the predicted geometric outputs and provide fitness information regarding individual material designs. Finally, we build an evolutionary algorithm, which, taking feedback from the machine learning model, guides the evolution of the design population and optimizes autonomously. During the optimization process, we impose physical constraints on the mutation methods to limit design complexity and assure converged designs are fabricable under real-world conditions.

This work demonstrates that we are able to explore the vast design space efficiently and learn important trade-offs and relationships of the parameter landscape.

154f Establishing Structure-Function Relationships in Low-Dimensional Mxenes Using Machine Learning

Tej Choksi, Pranav Roy, Lavie Rekhi, Hong Li, See Wee Koh

Two-dimensional materials like MXenes are compounds of transition metals and carbon/nitrogen. MXenes have exquisitely tuneable electronic properties leading to promising applications in energy storage/conversion and electromagnetic shielding. One such property is the work function of MXenes. Experimental studies indicate that the work function can be altered across a wide-range through changes to the MXene composition or altering the surface termination of MXenes. Recent advances have expanded the range of possible surface terminations to a variety of p-block elements like O*, N*, F*, Cl* S* etc. Another approach to tune the electronic properties of MXenes is by creating heterostructures of metal films on MXene sheets. The stability of these heterostructures is governed by the adhesion energy. This metric of stability determines whether a metal sheet prefers to form a low-dimensional heterostructure on the MXene or prefers to exist as 3D nanoparticles. This expanding space of materials presents a new challenge for efficiently predicting structure-function relationships.

To this end, we construct a series of machine learning models that predict functional properties of MXenes like the work function and the adhesion energy of metal overlayers on MXenes from a feature space consisting of physico-chemical properties of the constituting elements. The feature space includes properties like the electronegativity, ionization potential, orbital radii, lattice constants etc that are readily available in databases. Our machine learning models include ordinary linear regression, neural networks, and random forests. To lend physical interpretability to our non-linear models, we also employ symbolic regression. We employ two independent techniques to determine the most important features that govern these structure-property relationships. First, we perform a sensitivity analysis using permutation feature importance. Next, we compute the occurrence probability of different features in the most accurate models.

A 15-feature neural network model emerges as the best performing model and can predict the work function of MXenes from the properties of constituting atoms with a training and testing error of 0.13 eV and 0.25 eV respectively. The sensitivity analyses illustrates that the work function is more strongly dependent on the properties of the surface termination as against properties of metal cation or carbon/nitrogen anion further below the surface. Leveraging insights from our sensitivity analyses, we construct a simpler 5-feature neural network that predicts the work function of MXenes with errors of 0.28 eV, which is comparable to the best performing model. We implement a transfer learning approach that efficiently re-trains the neural network as new data points from additional first principles calculations are added. Experimental measurements of the work function of Ti_3C_2 terminated with F, O, and N atoms broadly confirm the trends in the work function established through our model.

While non-linear models are necessary to predict the work function of MXenes, the adhesion energies of metal films on MXenes can be robustly estimated using simple linear models. Key features that govern these adhesion energies include the electronegativity of the termination, the strain on the metal film, and the work function of the support. Model predicted adhesion energies are estimated with errors of 0.02 eV/A^2. These adhesion energies are used as inputs to determine the morphology of low dimensional metal/support heterostructures through a Winterbottom construction. Taken together, our study highlights the power of machine learning methods towards predicting the functional properties of two-dimensional materials from a feature space consisting of readily available physico-chemical properties. Furthermore, machine learning methods can engender physical-insights that help rationalize such structure-property relations.

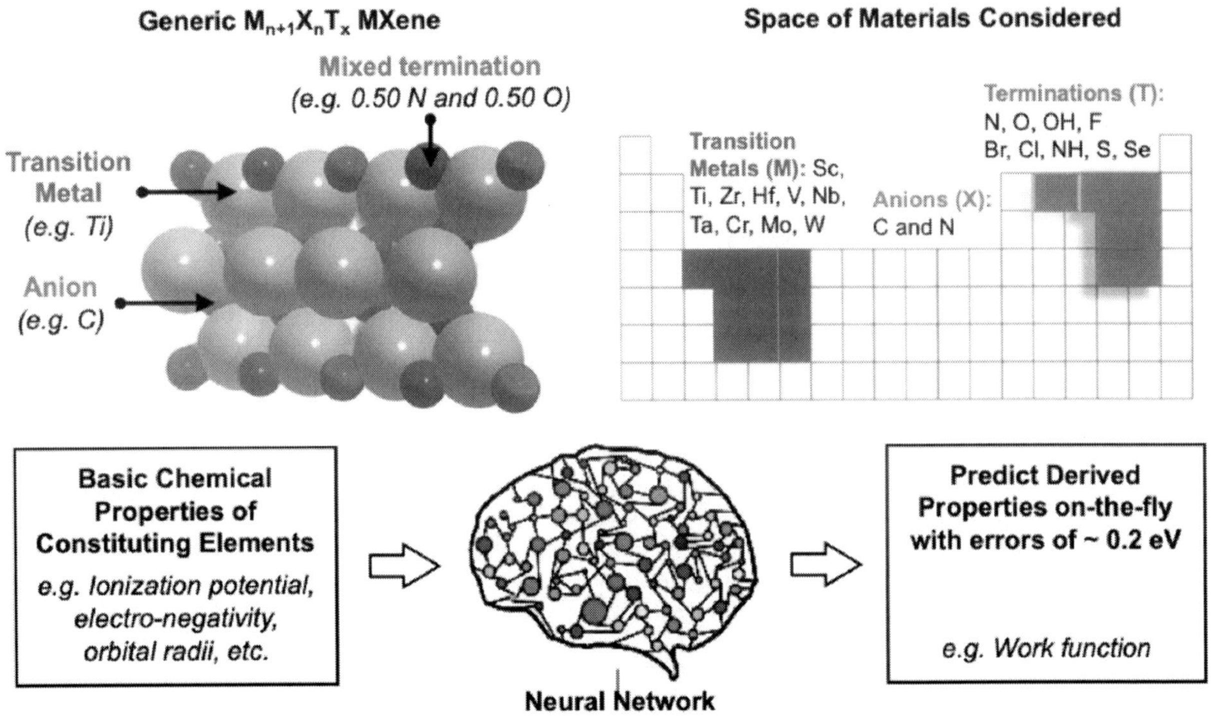
Generic $M_{n+1}X_nT_x$ MXene
Mixed termination
(e.g. 0.50 N and 0.50 O)
Transition Metal
(e.g. Ti)
Anion
(e.g. C)
Space of Materials Considered
Transition Metals (M): Sc, Ti, Zr, Hf, V, Nb, Ta, Cr, Mo, W
Anions (X): C and N
Terminations (T): N, O, OH, F
Br, Cl, NH, S, Se
Basic Chemical Properties of Constituting Elements
e.g. Ionization potential, electro-negativity, orbital radii, etc.
Neural Network
Predict Derived Properties on-the-fly with errors of ~ 0.2 eV
e.g. Work function

154g High Strength Polymer Composites Design for Lightweight Vehicles Via Stacked Transfer Learning and Explainable Artificial Intelligence

Jaewook Lee, En Sup Yoon, Dongil Shin

According to the global low-carbon policy, the need to develop lightweight materials for the purpose of improving vehicle fuel efficiency is increasing. Among them, polymer composites are used for about 12% of the interior and vehicle frame of vehicles, and the application parts are gradually expanding based on excellent moldability and lightness. Accordingly, it is required to develop a high-strength polymer composites that compensates for the disadvantages of plastics that are weak in strength.

When designing materials, the mechanical properties (tensile strength, ductility) of the polymer composites are important metrics, and the tendency of mechanical properties depending on the composition can be confirmed. However, the complex correlation according to the type of polymer and reinforcement is not clear. Accordingly, material designers rely only on experience and intuition to design polymer composites, which consumes a lot of time and workforce. Most of the previous studies used machine learning to predict the mechanical properties of polymers or reinforcements individually. In the case of the mechanical property prediction model of the polymer composites developed in the previous researches, The first study predicted the mechanical properties of the polymer composites according to the microstructure, and the mechanical properties of the composites according to the specific bonding structure can be predicted, but the mechanical properties of the composites according to the type of the polymer and the reinforcement cannot be predicted. The second study is the most closely related to this study as it developed AI for predicting mechanical properties of polymer composites according to the individual properties and synthesis ratio of polymers and reinforcements. The two-dimensional molecular structure information and individual mechanical properties of polymer and reinforcement, as well as process variables of polymer composites, are required for the model. Therefore, when designing a polymer composite using this prediction model, there is a disadvantage in that it is not possible to make predictions about polymers and reinforcements without data. Like most material design studies using machine learning, the amount of trainable data for polymer composites design is absolutely insufficient, so the coverage of material design has been very limited. In addition, all of the developed predictive models are black-box models and do not provide any judgment basis to users regardless of their accuracy.

In this study, we proposed an AI model for predicting mechanical properties of a polymer composites using only the three-dimensional molecular structure, synthesis ratio, and process conditions to cover a wide range of chemical space without requiring experimental values of the constituent materials. The problem of data shortage for most machine learning studies was overcome through transfer learning. In addition, LIME and SHAP, the most popular methods of the Explainable AI (XAI), were used to present a persuasive judgment basis to experts and non-experts in the field of polymer composites design.

First, A total of 265 types of polymer composites were produced to generate train data, which were prepared by adding fiber-based and inorganic fillers in 5 different compositions to 14 types of polymers most used in vehicles. Next, mechanical properties such as tensile strength, tensile modulus, elongation, Poisson's ratio, flexural strength, and flexural modulus were obtained through tensile and flexural tests. The first part of the model is a parallel connected CNNs that learn the representation of the three-dimensional molecular structure of polymers and reinforcements. But training CNNs requires large amounts of data, so in order to efficiently extract features such as structural information and functional group from molecular structure, pretrained CNNs were trained by mechanical property values such as tensile modulus, Poisson's ratio, flexural strength, and flexural modulus of polymer composites. After that, an FC layer was stacked to the pretrained CNNs and was trained to predict the tensile strength and ductility of the polymer composites through transfer learning with 265 types of polymer composites data. The performance of this predictive model based on 3D structures increased by approximately 7%, from 84% to 91%. Finally, we select 30 polymers and reinforcements suitable for vehicle lightweight materials and polymer composites composition was set between 90% and 50% as the search space. After that, we use the developed mechanical property prediction model and Bayesian optimization, high-strength and high-ductility polymer composites candidates were searched within the search space. The high-strength, high-ductility polymer composite resin design system was demonstrated five times, and the accuracy was about 87%, confirming the high accuracy. We also used an explainable A.I. (XAI) algorithm to reveal the relationship between three-dimensional molecular structure and mechanical property. Three-dimensional molecular structures that have a great influence on the model, such as CH3 and benzene ring, have been identified, and through this model analysis, a more reliable model can be provided to design experts and non-experts.

AUTHOR INDEX

AIChE
120 Wall Street, FL 23
New York, NY 10005-4020

ISBN 978-1-7138-7914-5

CURRAN ASSOCIATES INC.
proceedings.com